# INTRODUCTION TO CHEMICAL ENGINEERING THERMODYNAMICS